Physics of TRUTH
Science of peace and prosperity

Humanity is not aware yet that physics has all potentials to bring us peace and prosperity. The Physics of TRUTH has answers to the fundamental questions of today's natural and human sciences.

I dedicate this book to all who dare to build a new scientific paradigm which will fully integrate the most precious we can attain: "The TRUTH".

Thanks to Dr. Steve Kaufman for his fruitful discussion and financial support for the publication of this book.

Amrit Srecko Sorli
Foundations of Physics Institute
Planet Eart

Author's foreword:

"TRUTH" is the state of a realized enlightened human being who is free of all scientific, religious and cultural dogmas and convictions. Only a realized man can build an intercultural planetary society of prosperity, harmony and peace. Religions are dividing us, cultures are dividing us. Science enriched with consciousness is the only one which can bring us together.

The Physics of TRUTH is a new scientific paradigm based exclusively on an elementary perception and bijective function of a set theory. We can observe in the universe five fundamental elements: matter, energy, space, the run of changes and time. The one who observes these five elements is the observer which represents the sixth fundamental element. These six elements represent the set universe X:

$$X : \{M_x, E_X, S_X, C_X, T_X, O_X\}.$$

In Physics of TRUTH each of these elements is related with a bijective function of a set theory

$$f : X \rightarrow Y$$

with exactly one element in the set universe model Y:

2

$$Y : \{M_Y, E_Y, S_Y, C_Y, T_Y, O_Y\}.$$

Between the set universe X and the set model universe Y, there is a bijective function. The Physics of TRUTH is the beginning of a bijective epistemology in physics where no element in the set universe model Y can be introduced without being observed with elementary perception. Physics of TRUTH has proved that space has physical properties, namely, a variable energy density which generates mass and gravity. The energy of space is dark energy. A photon is the wave of space. Space is the medium which transports mass, gravity and electromagnetism.

No signal can move in time, because time is merely the duration of signal motion in space. CMBR cannot have an origin in some remote physical past which is non-existent. CMBR has its origin in space in which it is always NOW. Physics of TRUTH is the end of the Big Bang cosmology model.

In Physics of TRUTH, Einstein's idea of the completeness of a given model is fully realized and his visionary vision that only NOW exists:

...there is something essential about the NOW which is just outside the realm of science. People like us, who believe in physics, know that the distinction between the past, present and future is only a stubbornly persistent illusion.

Albert Einstein

The ground of future physics should be epistemologically where one is aware of the triangle: perception — mind elaboration — experience. We cannot build physics on mind elaboration which is not based on perception. Today, physics has lost contact with the physical world: in order to explain a given phenomenon physicists are just adding elements in the process of mind elaboration. Nobody cares if they do not have an origin in elementary perception. The result is that we have today several models which are not related; for example: the electromagnetic field, the Higgs' field and the gravitational field are three basic fields in today's physics and we are not able to join them in one model.

The bijective epistemology presented in this book is a new approach which solves all the fundamental puzzles of today's physics. Bijective epistemology is a developing physics which is 100% an exact picture of the physical world.

Table of contents:

1. What is time, what is space-time, what is space?

Let's start with the famous Albert Einstein's quote related to time: "...there is something essential about the NOW which is just outside the realm of science. People like us, who believe in physics, know that the distinction between the past, present and future is only a stubbornly persistent illusion."

Einstein experiences the distinction between the past, present and future as an illusion. Einstein is one of the greatest scientific minds ever, we cannot exclude that his statement might be right. This means that it might be true, that our experience of time as "past-present-future" is only an illusion of today's scientific mind. In Physics of TRUTH, time is what we measure with clocks. With clocks we measure the numerical order of material changes, i.e. motion, which runs in space. The smallest unit of the numerical order is Planck time, and the biggest unit is a year. In space it is always NOW. Linear time as "past-present-future" is a psychological time, which belongs to the mind. A common

7

observer in physics is experiencing a run of changes indirectly in the frame of psychological time, whereas a conscious observer is observing a run of changes directly, as they run in space, where it is always and only NOW.

In 1980 I started studying Einstein's Relativity Theory in detail and privately without a mentor. I trusted myself more than others. After a few years I discovered that in Special Relativity time there is not the 4th dimension of space. The space-time model is not 3D + T, the space-time model of Special Relativity is 4D. The so called temporal X4 dimension of space-time is spatial too. The formalism X4=ict confirms time t is not X4. Einstein was aware that time is not a physical dimension, which is why he put in the formula the imaginary number i. In Special Relativity, time is the numerical order of a photon's motion in space. Each Planck distance the photon passes through corresponds exactly to one Planck unit of time. The sum of Planck time is the duration of the photon's motion through the distance d between the source of light and its receiver.

In 2000, the Italian physicist Davide Fiscaletti contacted me and asked me for my cooperation on time research. In the following years we published together six articles on time and Special Relativity in the Canadian journal Physics Essays. We showed clearly in this work that the conviction of time being the 4th dimension of a space-time model is not valid and that the space-time model is not the most convenient picture of the fundamental arena of the universe. We proposed as a fundamental arena, a universe of 3D space, which has its origin in a 3D quantum vacuum, where fundamental time is merely a mathematical parameter, as a numerical order of material changes running in space. Fundamental time, when measured by the observer, gives emergent time, which is duration. The duration of a given material changes, i.e. motion in space is the sum of Planck times:

$$t = t_{P1} + t_{P2} + ... + t_{PN} = \sum_{i=1}^{N} t_{Pi} .$$

There is no duration without measurement from the side of the observer: changes run in the NOW. When changes are measured by the observer, their duration comes into existence. This fact confirms the importance of the observer in physics, which has been known since the 20th century: a given particle has its different super-positions, which exist in a quantum vacuum. The one which is observed becomes real.

The conviction of 20th century physics was that fields and particles exist in an empty space deprived of physical properties, which in physics is described with the model of space-time. Epistemologically this view is weak; it presumes that matter is primary and space is secondary. After solving the question on what is time, my vision was that space and matter are epistemologically equally positioned. This means that space has the concrete physical properties which are described by the "curvature" of the space-time model in General Relativity. A few years ago my idea was that the curvature of space is merely a mathematical description of "space variable energy density": in flat space energy, the density of space is at the maximum and is decreasing with the increasing of its curvature. Max

Planck metrics is the key to the formula which will express the idea presented above.

The energy density of "empty space" without material objects according to the Planck metrics is: $\rho_{PE} = \dfrac{m_p c^2}{V_P}$, where m_p is the Planck mass, V_P is the Planck volume and c is the speed of light. When this energy of space forms particles and solid bodies, its energy density in the area of a given particle or solid body diminishes according to the amount of their energy. This can be written in the following formula:

$$E = mc^2 = \Delta E_S = (\rho_{PE} - \rho_{SE}) \cdot V,$$

$$E = mc^2 = (\rho_{PE} - \rho_{SE}) \cdot V$$

where E is the energy of a given particle or solid object, m is mass of the particle, ΔE_S is the diminished energy of the space where the particle is situated, ρ_{PE} is the Planck energy density, ρ_{SE} is the energy density of space (quantum vacuum) in the centre of the particle or solid object, and V is the volume of a given particle or solid object.

I informed Fiscaletti of that formula, who liked it. We prepared an article on the subject but no journal with a science citation index was ready to take the article in the reviewing process. The editors informed us that, in physics, fields and particles exist in an empty space and that they were not open for discussion on this subject. We published our model of "Dynamic Quantum Vacuum" in some journals of lesser importance; and finally, in 2016, in the Ukrainian Journal of Physics, which is recognized by the European Physical Society. The ice was eventually broken.

In the physics of the 19th century, the origin of space was ether, and light was a vibration of ether. The American physicists Michelson and Morley carried out an experiment with an interferometer, which would confirm that light is a wave of ether. The experiment was designed with the premise that planet Earth moves in stationary ether. The outcome of the experiment was a big surprise for all physicists, namely, the experiment gave a null result. In order to "save" the prevailing "ether model" of physics, in which light is the wave of ether, the physicists Lorentz and FitzGerald introduced the idea of "length contraction" in the direction of motion. They explained the negative outcome of the Michelson-Morley experiment with the interferometer beam positioned in the direction of the Earth's motion, causing it to shrink slightly because of its motion. The Ether model was saved for a few years.

However, in the beginning of the 20th century ether was abolished from physics. In 1905, Einstein published his Special Relativity in which a photon moves in an empty space deprived of physical properties. The ideas of length contraction and empty space have no stable epistemological basis, and their adequacy with physical reality is questionable. This subject will be examined in detail further on.

According to the Big Bang Theory, universal space is expanding. The statement: "Space is expanding" is not falsifiable: we cannot prove it or disprove it, so it does not enter the realm of science. We can observe that distances between objects moving in space change, but out of this observation we cannot conclude that space is expanding or shrinking. NASA results confirm universal space has a Euclidean shape. This means universal space is infinite and "flat". The curvature of space in General Relativity is the mathematical description of its variable energy density. Cosmological models of a finite universe with a Riemann shape are not in correspondence with a physical universe in which space is infinite.

In the theory of infinite numbers, it has been proven by the Slovenian mathematician Ivan Vidav that when we say that the cardinal number of natural numbers is smaller than the cardinal number of real numbers, which we do not dispute. When we say that the cardinal number of natural numbers is equal to the cardinal number of real numbers, we also do not dispute. This means "infinity" in physics is not a metrical term. Infinite distance + 1,000 km = infinite distance. For a rational scientific mind, the size of the universe will always remain a mysterious secret. Albert Einstein used to say: "The most beautiful thing we can experience is the mysterious. It is the source of all true art and science". The rational

14

scientific mind cannot fully grasp infinity; the proper research tool for knowing the infinity is consciousness.

2. What is "relative" in the universe?

The Relativity Theory of the 20th century brought into physics a new understanding, namely, that time and space are "relative" in the sense that they can become longer or shorter, they can dilate or shrink. Physics of TRUTH has proved that time is merely the numerical order of changes running in space. This means that time is not a physical dimension in which changes run and cannot dilate or shrink. The conscious observer is directly experiencing (without the interference of psychological time) that clocks have different rates in different inertial systems. The GPS confirms that clocks run slower on the satellites than on the Earth's surface because of a Special Relativity effect by 7 microseconds and faster, because of a GR effect of 45 microseconds per day. In Physics of TRUTH (POT) this direct experience means that in the universe the rate of clocks changes with velocity, i.e. motion is relative and depends on the variable energy density of space, which originates from the quantum vacuum.

The GPS satellites are moving faster than the Earth's surface. Every material object in motion has its kinetic energy, which has its source in the quantum vacuum energy, which is additionally condensed in the moving object or particle. The energy density of the space (quantum vacuum) ρ_{SE} in an object with mass m and volume v diminishing in its centre according to the formula below:

$$\rho_{SE} = \rho_{PE} - \frac{m \cdot c^2 + EK}{V}.$$

The diminished energy density of the quantum vacuum causes clocks, and all material changes in motion, light included, to slow down. With the increasing energy density of the quantum vacuum, the rate of the clocks and the velocity of material changes become faster. The result is the General Relativity effect of the GPS system. In POT this effects are valid for all observers on the Earth surface, on the satellites in the airplanes.

In POT we do not have an "internal observer", "external observer", "length contraction" and "time dilation", which were models used by Einstein to mathematically describe the fact that light has the same speed in all inertial systems. In POT the photon is a wave of the quantum vacuum, in which all inertial systems, all observers, and all sources of light move. That's why light has the same speed in all inertial systems for all observers, and obeys the law of the Doppler effect. In POT, light is described in 3D Euclidean space, with a Galilean transformation for the spatial coordinates X, Y and Z, and the Selleri transformation for time t:

$$t' = \sqrt{1 - \frac{v^2}{c^2}} \cdot t .$$

The Selleri transformation shows that time is not linked to spatial coordinates, it is linked only to velocity, which confirms that time and space are separate phenomena.

In POT we do not have "gravitational time dilation" as in General Relativity (GR). Shapiro measured that light has a minimally smaller speed when moving in strong gravity. This result was than interpreted as a dilation of time, as a 4th temporal dimension of space, in which light moves. In POT, stronger gravity corresponds with a stronger curvature of space and a lower energy density of the quantum vacuum. We know in physics that a lower density of the medium diminishes the propagation of the signal moving through it. The same is the case with the Shapiro effect. The lower energy density of the quantum vacuum changes its permittivity and permeability, which causes the speed of light to diminish slightly.

The idea that the speed of light could have the same small variations has been strictly excluded in physics since 1905. Minimal variations of the speed of light keep Special Relativity **(SR)** and **GR** safe. These two theories are the pillars of physics, which is the house of science on stable ground. In my view, physics can only minimally improve the existing models of the 20th century physics with profound epistemological research and small "adjustments" and improvements of the existing models. However, POT is developing and unifying **SR** and **GR** by introducing a variable energy density for space, which originates from the quantum vacuum, in giving them a more solid physical ground and more explanatory power, as we will see in the next chapters.

3. What is the origin of light?

In POT, the fundamental research tool is "bijective epistemology". In bijective epistemology each element of the model corresponds exactly to one element in the physical universe. Bijective epistemology is based on the bijective function of a set theory. I came to this idea because, in the model of physics, we sometimes have elements for which we do not know if there are correspondent elements in the physical universe. Bijective epistemology proves, for example, that the element of space-time, where time is considered to be the 4th physical dimension of space, has no counterpart in the physical universe. According to bijective epistemology the term "empty space" also has no correspondence in this physical world. Space in which matter exists has concrete physical properties. In POT, a photon is a wave of space which originates from a 3D quantum vacuum. Combining two well-known formulas of physics, we see a photon has energy E and so also mass m:

$$E=mc^2$$

$$E = h\nu$$

$$mc^2 = h\nu$$

$$m=\frac{h\nu}{c^2} \quad (8),$$

where h is the Planck constant and ν is the photon frequency.

The velocity of light has minimal deviations, which depend on the energy density of the quantum vacuum as it changes its permittivity and permeability:

$$c = \frac{1}{\sqrt{\mu_0 \varepsilon_0}} .$$

The velocity of light is constant in all inertial systems moving in a quantum vacuum with a stable energy density. The only parameters that influence the speed of light are the variable energy density of quantum vacuum, and the dragging effect of the quantum vacuum. The Earth's rotation also causes the quantum vacuum around the Earth to rotate. When we measure the time of a light signal moving from A to B on the Earth's surface in the direction of the Earth's motion, the time t is shorter than when we measure light moving from B to A. This is because when light moves from B to A, the medium which carries the light, is moving from A to B. The same happens when you swim in a river upstream, or you swim downstream. **POT** fully explains the Sagnac effect, which **SR** cannot explain when a photon there is said to move in empty space.

4. What is the origin of energy, mass and gravity?

To find the right model describing the relation between energy, mass and gravity we have to follow the methodology which is epistemologically correct. Let's describe an epistemologically correct methodological procedure in three steps with the simplest example of uniform motion:

1. Uniform motion is observed.

2. A mathematical model is built to describe the uniform motion: $d = v \cdot t$.

3. Experiments have been done which confirm that the mathematical model is right.

In today's physics, the most important thing is the mathematical model. If you have an article with a good mathematical model you can publish it in the SCI (Science Citation Index) journal. Observation and experimentation are not required. The classical example of such a theory without observation and experimentation is "The

String Theory", which is a pure mathematical theory. Also **GR** is a mathematical theory whose beauty is that it exactly describes planetary precession and other effects of **GR**. POT gives **GR** physical meaning. The curvature of space in **GR** is in POT, the mathematical description of space energy density.

In today's physics we have three main fields: the electromagnetic field, Higgs field and the gravitational field. In POT we have only one field, which is the quantum vacuum itself. In POT, the concept of mass has three aspects: mass m of a given particle or solid body, as the amount of quantum vacuum energy incorporated in a given particle or solid body; Inertial mass m_i, as the physical property of a given solid body to have resistance when we put it in motion, and to have inertia to remain in motion; and Gravitational mass m_g as the physical property of a given solid body to attract all other solid bodies around it. All these three aspects have their same origin in the variable energy density of the quantum vacuum, so we can write following formulae:

$$m = m_i = m_g$$

$$m = \frac{(\rho_{PE} - \rho_{qv}) \cdot V}{c^2},$$

where ρ_{PE} is the Planck energy density, ρ_{qvE} is the energy density of the quantum vacuum (space) in the centre of a given solid body or particle, and V is the volume of solid body or particle.

POT functions well without the Higgs' boson and without the graviton. In **GR,** gravity is thought to propagate between solid bodies. The carrier of gravity is a theoretical particle, the graviton, which moves at the speed of light. The epistemological weakness of the graviton is that when it moves from body A to body B it has no physical contact with them, and cannot keep them together. In POT, solid bodies and particles exist in the quantum vacuum, and diminish its energy density accordingly to the amount of their energy. The diminished energy density of the quantum vacuum creates a pressure from the outer quantum vacuum, which pushes towards the quantum vacuum area with a lower energy density. This pressure of the outer quantum vacuum is gravity.

In the original version of **GR,** gravity is carried by the curvature of space. Einstein saw it only as a mathematical description, where gravity has a physical origin. His idea of gravitons, which are similar to photons, was that gravitons carry gravity and photons carry electromagnetism. This was a theoretical mistake, which inspired physicists for the next hundred years. The graviton has not been discovered yet, and will not be discovered because it does not exist in the physical world.

Einstein developed a mathematical model according to which gravitational waves (GW) dilate or shrink space. On the basis of this model, LIGO has developed a theory according to which GW could dilate or shrink the beams of an interferometer. Recently it was discovered that laser light in the LIGO interferometer has a variable speed, which is explained by the variable length of the interferometer beams caused by GW. This discovery has an epistemological weakness, namely, the variability of the lengths of the interferometer beams was measured indirectly. It could also be that GW are changing the energy density of the quantum vacuum, and so it could be that its permittivity and permeability causes the variability of the speed of light. This explanation is more realistic because it is difficult to imagine that GW, which are extremely weak, could change the length of 4,000m long beams set in solid reinforced concrete.

In today's physics the most important element is the mathematical model. Once the mathematical model is established, physicists can start designing experiments to confirm that the mathematical model is correct. The "discovery" of GW is a classic example of this inappropriate research methodology.

By positing the "discovery" of GW, step 1 is missing. GW have never been observed. The procedure started with step 2, and the experiments of step 3 do not prove the existence of GW, they simply prove that the speed of the laser signal is variable. The cause of the variable speed of the laser can be explained by the GW changing the lengths of the beams. However, our research group has an alternative explanation, namely, the GW are changing the energy density of the quantum vacuum, and its permittivity and permeability causes the variability of the speed of the laser.

The same methodology has been used in the "discovery" of the Higgs boson. The idea that some field gives mass to elementary particles was put into mathematical form by Peter Higgs. Then the CERN designed experiments to prove the existence of a particle which carries the field responsible for the mass of elementary particles. In POT, the Higgs boson is the characteristic flux of quantum vacuum energy released by the collisions of two protons. It does not prove the existence of the Higgs field, which has no existence in physical reality. As with the graviton, the Higgs field is the wrong theoretical model. Great minds are creating great ideas and also big mistakes: Lorentz' idea of length contraction, Einstein's idea of the graviton, Higgs idea of the Higgs field, Feynman's idea that the positron is an electron moving backwards in space-time. Every particle and solid body can move in space only, and time is the duration of their motion.

The idea of time travel, which has stirred human minds for a century, is categorically excluded by POT. We can only move in space, and time is the duration of our motion.

The English Physicist Brian Cox thinks we cannot travel to the past; we can only travel to the future. Time travel to the past creates a contradiction; one could travel to the past and kill one's grandfather, and so could not be born. In order to prevent this contradiction, Stephen Hawking introduced "the chronology protection conjecture", that the laws of physics prevent time travel on all but sub-microscopic scales. According to **POT** there is no need for "chronology protection" because from the micro to the macro, scale motion happens only in space and never in time. The idea of time travel was raised when Kurt Gödel discovered in 1949 that a closed time-like curve (CTC) can exist. A CTC is a world line in a Lorentzian manifold, of a material particle in space-time that is "closed", returning to its starting point. Back in 1949 the conviction was that the 4th dimension of space-time *is* time, and so travelling back on the CTS one could travel into the past. In POT we know that the 4th dimension of space-time is spatial too, which means that travelling back into CTC one will arrive at the place of departure and not in some past. We always travel in space only where it is always NOW.

In POT, gravity, mass and electromagnetism are fully integrated. In today's physics, the gravitational field, the Higgs field and the electromagnetic field are not fully connected because they represent partial models which do not have the bijective epistemological correspondence of POT. Also in today's physics the observer is not the consistent part of the existing models, whereas POT fully integrates the observer.

5. What is the origin of the observer?

Back in 1987, in Ljubljana I was measuring, in different labs, the eventual difference of mass between living and dead worms (lumbricus teresticius). I was inspired by the bio-photon research of the German biophysicist Albert Fritz Popp, who was searching for bio-photons in living organisms. The results of the Polish scientist Slawinski confirm that bio-photon radiation from dying organisms is 10 to 100 times greater than the radiation from organisms in good health. I knew that this bio-photon radiation would be impossible to measure as a difference of mass; my intention was to measure the eventual correlation between the quantum vacuum and life. It could be that in the living organism beside bio-photons are some unknown structures of the quantum vacuum that have measurable mass. Preliminary results showed that 1 gram of dead worms have about 5 micrograms less mass than 1 gram of living worms.

During the experiments, which required a very precise procedure, I became aware that I observed the experiment, I wrote down the measurements, and I did not know the observer. This was quite a big surprise for me, and so I started studying psychology intensely. After a year I hadn't found any answers, and so I turned my attention to Buddhist psychology where the observer is the central point of the research. I wrote a few articles on the relationship between the observer in Western science and the observer in Buddhism. When it was clear to me that knowing the observer is not in the domain of the human mind I started regularly practising the Buddhist meditation Vipassana. The more I became aware of how my mind functioned, the more I entered in the state of the conscious observer, who is aware of psychological time. I read the book of Eckhart Tolle "The Power of Now". I sent him a few letters where I wrote that his teaching had a scientific basis in the right understanding of time as a numerical order of changes occurring in space where it is always NOW. He never answered me.

The observer is the common element of all branches of science, and has immense potential to develop a so called "Theory of Everything" (TOE) where the observer is the central part of the research. As long as we do not know the observer, all our knowledge is used for egoistic partial interests, which creates the situation we face today on our planet. The observer cannot be measured by instruments, but it can be experienced. I believe that reductionist science, where things only exist which we can measure, has reduced human experience enormously in this beautiful universe. The TRUTH is beyond the rational human mind, which can only point towards it. The TRUTH can be known only by being experienced. In POT, the self-experience of the observer has full scientific validity despite the fact that it is not measurable. In POT, the validity of the experience is equal to the measurement. What we measure is real, what we experience is real too.

We cannot build a scientific model of consciousness in bijective epistemological correspondence with consciousness itself, in the way we can do with the "material universe". Consciousness is beyond the range of the rational mind. I have developed a model where consciousness is described as the n-dimensional Hilbert space. The mind represents fewer dimensional Hilbert spaces, which then follows the 3D quantum vacuum of material objects, which are its structures. If the mind had its origin in the neurons of the brain, which are made out of molecules and atoms, the mind would not be able to develop n-dimensional geometries. Its geometry would end with the 3D Euclidean geometry. If the mind did not have its origin in consciousness, the mind would not be able to mathematically formulate infinity. Mathematics is the purest manifestation of consciousness in the rational realms of the human mind.

POT results confirm that consciousness is the same in every human being, and that the origin of the observer in every human being is the same consciousness. Consciousness acts via the mind and bio-photons on the microtubules of the brain.

It is of essential importance that we start teaching POT, and especially its part in the origin of the observer. The experiential science of the observer has an immense potential for the integration of different cultures and religions. Modern society needs the appropriate "spiritual education" in order to overcome the problems we have been facing since the beginnings of human civilization: wars, violence, starvation.....

In POT the photon is a wave of universal space corresponding to the fundamental "3D quantum vacuum" or "3D physical vacuum". In quantum mechanics the energy of the photon is:

$$E = n \cdot h \cdot v,$$

Where h is the Planck constant, v is the photon frequency and n is an integer number (1,2,3...). Consciousness can be described as the photon which has infinite frequency and exists in n-dimensional Hilbert space:

$$\lim_{v \to \infty} nhv = Consciousness$$

In the formula above the integer n represents the dimensionality of Hilbert spaces. When vibration v becomes infinite, the integer n becomes the cardinal number of the natural numbers. Consciousness is the vibration of the n-dimensional Hilbert space whose limit tends to the infinite value of frequency v and zero value of the wave length λ. From that, it follows that the velocity v of

36

consciousness is zero:

$$v_C = v \cdot \lambda = 0.$$

This mathematical model cannot be considered as a real picture of consciousness; it only indicates its real nature. Consciousness is beyond logic and therefore beyond mathematics, which can only help us to build an approximate model. Mathematics cannot explain the physical reality of the universe; it can only describe it with its limitations. Consciousness is a subject which can only be researched and known by experience.

I speculate that the logical part of the human mind has its origin in higher dimensional Hilbert spaces, where consciousness is directly minimally diminishing or increasing the energy density of the appropriate Hilbert spaces. Increasing energy density means 1 and decreasing means 0. In lower dimensional Hilbert spaces this binary mathematics is transformed into the decimal system. Sure, this is pure speculation, on which we cannot build a consistent model having bijective correspondence with the real world of the consciousness and the mind. I believe that a mathematical description of consciousness and the mind has limitations, and that "watching", "witnessing" the mind is the most appropriate research method in the field of consciousness and mind research. The more we are able to watch the mind, the more we are consciousness itself.

Higher dimensional Hilbert spaces are the "guiding waves" of De Broglie's "pilot wave theory". POT sympathizes with De Broglie and David Bohm, namely, that elementary particles are guided and accompanied by "pilot waves". Each particle has its correspondent duplicates in higher dimensional Hilbert spaces. In this sense, the whole universe is guided by consciousness. When the human mind is tuned with consciousness, it creates beauty and wealth. When the mind is disconnected from consciousness, it creates chaos.

6. What is the origin of life?

In POT, consciousness is the origin of life, which all over the universe has the tendency to develop in intelligent organisms. Humans do not have "free will". We have to be aware we are only the segment of the large universal process in which matter is developing to consciousness. Our real will is the will of the universe, which is that we develop towards consciousness.

The physical homogeneity of universal space also includes biological homogeneity. Organic molecules for development of life have been found in the entire observable universal space. Universal space, which originates from the quantum vacuum, is syntropic energy. Matter is entropic energy. Life, as a process where entropy is continuously diminishing, is possible because matter exists in the syntropic energy of space. In POT the energy of space is missing "dark energy"; this means in the universe around 95% of energy is syntropic, and only 5% of energy is entropic. The self-organization of matter, which is today leading evolutionary theory, is possible because matter exists in the syntopic energy of space.

In POT, the human being and human society are subsystems of the universe. POT is surpassing geo-centrism and is introducing in anthropology, psychology, sociology and economy, the "cosmological approach", where human society is the subsystem of nature, which is the subsystem of planet Earth, which is the subsystem of the solar system and so on. Today, human society does not respect the fundamental laws of the universe, which results in the enormous problems of today's human society.

We teach our children that they live in a certain place, which has its own history. We teach them our religions; we teach them our cultures. We do not teach them that their real home is the universe. When they become adult their natural connection with consciousness is broken, and they become "civilized". History teaches us that every civilization that lost its connection with consciousness became lost in the labyrinths of their collective mind. The mind without consciousness is like a blind man seeking for an oasis in the desert.

The origin of life is hidden deeply in every human being, the animal in every existing thing. Eating animals is, according to POT, one of the main reasons for wars on the planet. It shows zero compassion and the immense cruelty of today's human beings. The flesh of dead bodies is dead without bio-photons and brings death. Plants are alive and full of bio-photons, which are an "information bridge" between the consciousness and the human body. A mind full of wrong ideas, which are not in concordance with the universe, is sick and is preventing the organism to live in harmony with consciousness. When the mind "takes over" we get psychosomatic diseases. When the mind is connected with consciousness, the organism is radiant with bio-photons, which means health. Bio-photons are the main information carrier between the cells and different organs. When this central information system does not work well, one will get sick.

People eating a lot of meat are lowering the power of their bio-photonic field, which is the bridge between the brain and consciousness. The brains of vegetarians probably function better than those of people eating meat. Pythagoras, Buddha, Voltaire, Mahatma Gandhi, Albert Einstein, Nicola Tesla, Leo Tolstoy, George Bernard Shaw, Leonardo Da Vinci, and many others genius were vegetarian.

In POT **a** deep connection with consciousness represents the basis for psychophysical health. When we are anchored in consciousness, our mind and body function well. POT validates the following equation:

$$health = consciousness,$$

which is valid both on an individual level and on the level of the entire human society. The more the given society is conscious, the more society is healthy, in the sense of living harmoniously with nature, prosperity and wealth. Bhutan is a living example of such a conscious and healthy society.

7. What is the origin of the universe?

This is one of the most frequently asked questions of the human mind since the beginning of human civilization. According to **POT** the universe has no origin. On the contrary, the universe is the origin of all things. The Universe has no beginning and no end; it is a system in permanent dynamic equilibrium. In black holes, the energy density of the quantum vacuum is so low that atoms become unstable, and matter transforms back into the energy of the quantum vacuum. In intergalactic space, the energy density of the quantum vacuum is at its maximum. The energy of the quantum vacuum is continuously transforming into cosmic waves, which build elementary particles. This circulation of energy is permanent we can say eternal. This eternity is NOW. The universe exists in an n-dimensional Hilbert space, which is consciousness. Linear time "past-present-future" enters into existence with the development of the human mind, and belongs to the human mind.

The idea that the universe was born more than 13 billion years ago (where time is the physical component of space-time, as the fundamental arena of the universe) out of an infinitesimal point of energy requires the existence of God. The father of the Big Bang Theory was the Belgian physicist Georges Lemaitre, who was a Christian priest. POT means the end of the Big Bang Theory (BBT) because no signal can move in time. All signals are only moving in space (where it is always NOW), and time is the duration of their motion. In POT, cosmic microwave background radiation (CMBR), which is the main proof for the Big Bang Theory, is the radiation coming from the quantum vacuum.

Also the "inflation phase" of the BBT is contrary the logic of **POT**. Stephen Hawking explains the inflation in his book "The Brief History of Time" by the multiplication of the energy of matter, which is positive, and the multiplication of the energy of space (gravitational energy), which is negative. The sum of both energies in the universe is always zero in the same ways as:

$$1 + (-1) = 0$$

$$2 + (-2) = 0$$

$$n + (-n) = 0$$

What is valid in mathematics is not valid in the physical world. Mathematics cannot explain physical laws; it can only describe them within its limitations.

We know today that around 70% of the "red shift" belongs to the gravitational red shift when light is pulling out of the strong gravity fields of the galaxies, and so loses some energy in the form of obtaining a lower frequency.

The POT cosmological model is based on bijective epistemology, direct experience and experimental data. Direct experience means that we do not interpret any more what we perceive in the senses. For example, we measure in the GW experiments of LIGO that laser light changes speed. POT gives an immediate explanation of this observation: laser light is variable because of the variability of the energy density of the quantum vacuum *caused* by the GW. The official explanation is that the GW change the length of the beams, which is not observed. In POT the scientific mind has lost its privilege to interpret experimental data. The mind has only the right to build a model to directly explain the experimental data. In POT there is no "epistemological gap" between the model and physical reality. POT is also in front of other models because it fully recognizes that changes run in the NOW. This insight is the basis for the development of physics and science in general.

Questionable interpretations of experimental data include the "discovery" that binary stars diminish in rotational time because of the emission of gravitational waves. In 1974, the American physicist J. H. Taylor, along with his research group, observed a binary neutron star called PSR B1913+16. They noticed that the rotational speed of binary stars around their axes diminishes over time. This is a fascinating observation and result, yet their interpretation of the data is questionable. They attributed the decrease in the rotational speed to a reduction in the binary stars' masses. POT considers this interpretation is valid; however, the reduced masses were considered to be due to gravitational radiation, which requires the existence of gravitational waves which have never yet been directly observed. In POT, gravitational waves might exist only as a variable energy density of the quantum vacuum, which propagates in outer space from stellar objects (binary stars, black holes), in which matter is transforming back into the energy of the quantum vacuum.

In POT, the diminished energy density of the quantum vacuum around stellar objects is also rotating. We call this the "dragging effect". In 2011, my idea was that this dragging effect is the physical basis for planetary precession when moving around the Sun. I informed Fiscaletti about this model, also sending him the basic formulas for variable energy density, and in a few months he developed a mathematical model on which basis he calculated planetary precession. He got exactly the same results as Einstein. After five years we managed to finally publish these results. Getting exactly the same results by using a completely different model was clear proof that our model corresponded to the physical reality.

Let's see how in the POT bijective epistemology is developed. The Merriam-Webster dictionary contains the following definition of epistemology: *"the study or a theory of the nature and grounds of knowledge especially with reference to its limits and validity."* In this chapter, we will examine how the foundations of physics are built. When searching for the foundations of physics it is important to look for an answer to this question: "Which fundamental elements of the universe are also the fundamental elements of physics?". Let's put ourselves in the position of an observer and see what are the fundamental elements of the universe. The first element is the observer himself. The second element is matter, the third element is energy, the fourth element is space in which matter and energy exist. Matter and energy are permanently changing, so the fifth element is change. A given change has its appropriate duration, namely, time as the sixth fundamental

48

element of the universe: Observer, Matter, Energy, Space, Change and Time.

From Einstein we know that matter and energy are made out of the same "stuff"; matter can transform into energy and, oppositely, energy can be transformed into matter. In physics, we call this intrinsic relationship between matter and energy the mass-energy equivalence principle, and it is expressed by the famous equation: $E = mc^2$. Matter and energy both exist in space.

During the 20th century, there was a conviction in physics that space is "empty" and without any physical property. In the General Theory of Relativity (GR), Einstein described gravity with a curvature of space. Einstein geometrized gravity. We will enquire now about the idea that space has only geometrical properties, and is without physical properties, from the point of view of epistemology. Let's conduct an experiment: you observe a material object in front of you. Photons are reflected from that object, and they come into your eye, where they are transformed into an electrical signal, which moves through the nerves to the centre of the brain, where the observer experiences the object. Your mind will elaborate on the information about the object which has translated by the eye into signals, and so the experience will not be exactly the same as the perception. In epistemology, we call that an elaboration gap between perception and experience:

perception ::: elaboration (gap) :::experience.

We need to answer the question: "Could the experience of this material object happen in an empty space which has no physical attributes. Can matter and energy exist in a space which only has a mathematical existence?". The idea of "empty space" as a fundamental arena is not falsifiable. The sentence: "Matter and energy exist in an empty space" is not falsifiable, which means it has no scientific value. It belongs to the philosophy of physics. In this book, we will establish a clear line between the philosophy of physics and the foundations of physics. We observe in physics that particles and antiparticles are continuously appearing from space and disappearing back into space. This experimental data shows that space has *physical properties* since, otherwise, *physical particles* could not appear from space and not disappear back into space. We can conclude that the physical elements of the universe, namely, matter and energy exist in space, which is also physical; it has its physical properties, it is a kind of "energy".

What are the physical properties of space energy? To answer this question, we will use Planck units, which belong to the foundations of physics. Planck energy density $\rho_{PE} = \dfrac{m_p \cdot c^2}{V_P}$ is the average energy density of an *empty* universal space in which there are no material objects. Out of this fundamental energy pool of the universe, particles and antiparticles continuously appear and disappear. This picture in which matter and energy exist in universal space, which is a universal pool of energy, is *organic* and rational. We can now rearrange our model of the universe in which we have four elements: Observer, Change, Time and Energy. To put this in a more mathematical form, we will call the universe set X which

has three elements and one energy subset
EX:

$$X : \{O_x, C_x, T_x\{EX\}\}$$
$$EX : \{E_x, M_x, S_x\},$$

O_x - observer, C_x - change, T_x - time, E_x - energy, M_x - matter, S_x - space (7).

In this picture, the appearance of particles and fields in the universe is no longer mysterious. The particles and fields have their origin in the energy of space as we have seen in chapter 6.

We see that the energy of space is somehow fundamental. All the different types of energy in the universe, including matter, are different forms of space energy. We can mathematically describe the universe as set X with three elements, a space energy SE_x subset which has two elements:

$$X : \{O_x, C_x, T_x\{SE_X\}\}$$
$$SE_x : \{E_x, M_x\}.$$

The mass-energy equivalence principle allows us to see matter and energy as one physical entity. We can describe the universe as set X with four elements: Observer, Change, Time and Energy of space which has forms of energy and matter.

The mass-energy equivalence principle allows us to see matter and energy as one physical entity. We can describe the universe as set X with four elements: Observer, Change, Time and Energy of space, which has forms of energy and matter.

$$X : \{O_x, C_x, T_x, SE_x\}$$

Space energy SE_x is the "missing" 95% of the energy of the universe (dark energy) which we also call a "quantum vacuum".

POT introduces the energy of space as a fundamental energy of the universe on the following experimental data:

- the conscious observer directly experiences that light has the same speed in all inertial systems.

53

- the conscious observer directly experiences the Doppler effect by light in all inertial systems.

The only reasonable conclusion is that light is the wave of the medium in which the source of light and all inertial systems move. This means universal space is not "empty". Bijective epistemology fully confirms this conclusion. In POT, the constancy of the speed of light is not a postulate as it is in **SR**; the constancy of the speed of light is the experimental data which allows POT to introduce space as the fundamental energy of the universe. As universal space is infinity, so also the amount of energy of space is infinite: $SE_x = \infty$.

In **SR** we have length contraction, which leads to a contradiction when we have, in a given inertial system moving with velocity v, two identical photon clocks. One is positioned vertically and other is positioned horizontally in the direction of the motion. At high velocity, v the horizontal photon clock will shrink, and so a photon will have a shorter path; so for the stationary observer the horizontal clock will run faster than a vertical photon clock. In POT we do not have length contraction, so there is no contradiction. All photon clocks in a given inertial system have the same velocity for all observers.

With the discovery of the final Riemann geometry, cosmologists started to use Riemann's finite spherical space as the model for universal space. For the rational scientific mind this model was most welcome because a rational mind likes finite things; infinity is beyond its comprehension. In POT, universal space is infinite. We can only study how the universe functions in observable space and also predict that the unobservable universal space behaves in the same manner.

Not only the infinity of universal space, but also the origin of the observer reaches beyond human rationality, and can be known only by experience. POT includes experiential research on the origin of the observer, so expanding the existent paradigm of physics: "Real is what we can measure and what we can experience".

In POT, every particle is structured energy of the quantum vacuum, which moves in a quantum vacuum. The motion of each particle creates a wave of vacuum. The particle and its correspondent wave are inseparable, and we have to examine them together.

The right imagination is the source of new discoveries and new knowledge. POT has shown that the imagination of time as being the 4th dimension of space is not adequate. From this insight the idea of the variable energy density of space has developed. By introducing space as the fundamental energy of the universe, POT has given physics new perspectives. If this research will continue it might give us new ideas on how to get free energy from space and how to use the energy of space for interstellar travel.

In POT, space is the direct information medium between entangled particles. Photons are the fastest information mediums in the universe, which need time (duration) to travel from object A to object B in space. Space. In my opinion, we do not know exactly yet in which n-dimensional Hilbert space that entanglement takes place. Hilbert spaces are also direct information mediums in remote viewing and telepathy. With the understanding that information does not run in time, but runs in space only, entanglement can be fully understood.

The Hilbert spaces of POT correspond to the David Bohm idea of the "wholeness" and the "implicate order" of the universe. In POT, wholeness means that matter is the vibration of consciousness in the lower frequencies of 3 dimensional space. "Implicate order" means that the behaviour of matter is guided by the higher dimensional Hilbert spaces. The model of Hilbert spaces also corresponds to Einstein's idea of the "hidden variables" which define the behaviour of elementary particles. Einstein, Podolsky, and Rosen argued that "elements of reality" (hidden variables) must be added to quantum mechanics to explain entanglement without action at a distance. In POT, 3 dimensional particles exist in the higher dimensional Hilbert spaces, which are the direct information medium between particles. In the EPR experiment, when we measure the spin of a particle A, the particle B will automatically get the opposite spin because higher dimensional spaces have the intrinsic function of keeping symmetry (in the sense of left or right spin) between entangled particles. Symmetry is the fundamental law of the universe, which is expressed in the "energy-mass-gravity" formula:

$$m \cdot c^2 = \Delta E_{Space} = (\rho_{PE} - \rho_{SE}) \cdot V$$

Mass *m* *of* a given particle (in the sense of the amount of energy) is symmetric to the loss of energy of space ΔE_{space} . The loss of the energy of space is symmetric to the difference between the Planck energy density ρ_{PE} and the energy density of the space in the centre of the particle ρ_{SE} with the volume *v* . The amount of mass of a given particle or solid body, and the corresponding diminished energy density of space as the quantum vacuum, represent the fundamental symmetry of the universe, which generates inertial mass and gravitational mass. In different articles we use this formula in a slightly different form, namely, the energy of space is called the "energy of the quantum vacuum".

8.   Epistemology Crisis of Today's Physics

In today physics it often happens that experimental data is interpreted as the proof of the phenomenon that has not been directly observed and is based only on the theoretical model. With the obtained data, the model becomes recognized as "real" and the phenomenon that the model describes also become recognized as "real" – it is acknowledged as the physical reality, although it has not been observed by instruments or human senses. This situation is leading physics into deep epistemology crisis, which is not yet seen today. The aim of this article is stressing the situation and finding the solution to overcome the crisis.

Recently, two Nobel prizes have been given for the discovery of phenomena which have not been observed by instruments or human senses, namely: Higgs field and gravitational waves. In the Higgs field research, it was measured that extremely rarely (one in milliard collisions of protons) is measured a characteristic flux of energy named "Higgs boson". The discovery of the

Higgs boson in today physics means the proof for the existence of Higgs field which was not measured or observed by human senses. In gravitational waves research, it was measured that the laser light motion in the LIGO interferometer needs sometimes (when gravitational wave is supposed to pass the interferometer) a bit longer or shorter time when passing the beams. The minimal time variability of the laser light is, in today physics, the proof for the existence of gravitational waves which were not measured or observed by human senses. This "epistemological gap" between obtained data and their interpretation represents a serious problem from the view of the epistemology of physics.

Special Theory of Relativity (STR), published in 1905, has deeply changed the methodology of physics. STR has caused that still today it is fully acknowledged that time is the 4th dimension of space which has physical existence. The formalism $X4 = ict$ has convinced the majority of physicists that time is the 4th dimension of space-time model. Later on, the acceptation of the space-time model has convinced physicists that time as the 4th dimension of space has real physical existence, although there is no experimental evidence for this and, moreover, nobody has ever seen time as the 4th dimension of space. With STR mathematical model it has happened that mathematics has overruled physics. If today you have a consistent mathematical model, every peer review journal will publish it, without bothering about the epistemological stability of the article. Epistemological stability means the level of the adequacy of the model with physical reality. The most epistemologically stable would be the model which is related to the physical reality with the bijective function of set theory. The bijective function prescribes to each element a physical reality as exactly one element in the model:

$$f : X \rightarrow Y \quad (1).$$

Recent research where the so-called "bijective epistemology" was applied has confirmed that the model of space-time has no "bijective epistemological stability". Time is not the 4th physical dimension of space

(Fiscaletti and Sorli, 2015). It seems that Einstein was aware of the fact that time has an exclusively mathematical existence and he added the imaginary number i in the formula $X4 = ict$. After a certain time i was removed from the formula, which then became $X4 = ct$. At the end constant c was written as 1 and formula has taken the form $X4 = t$, which was then the exact mathematical description of the conviction that material changes are running in some real physical time as the 4th dimension of space. This conviction has no epistemological stability, it is the biggest theoretical failure of the 20th century physics and, however, it is still valid in the mainstream physics today.

The weak point of today's physics methodology is that, based on mathematical models which are proved by *indirect experiments* (indirect experiments are the ones which do not directly measure the phenomenon that is the subject of a given research), the conclusions are taken about the existence of these phenomena as if they had an actual physical existence. The best examples are the existence of Higgs field and gravitational waves, which both have not been directly observed and measured. They have been only theoretically predicted and described with the mathematical model, for which we do not know how much it correspondents to the physical reality.

Length contraction was postulated by George FitzGerald (1889) and Hendrik Antoon Lorentz (1892) to explain the negative outcome of the Michelson–Morley experiment and to rescue the hypothesis of the stationary aether (Lorentz–FitzGerald contraction hypothesis) (FitzGerald, George Francis, 1889). Length contraction was later adopted by Einstein and used in his STR and also in his General Relativity Theory (GRT).

The idea that the length of a given moving object could change its length has no minimal common sense. When the idea was mathematically described, it has become plausible because, at that time, nobody did care much about the negative epistemological consequences of such a methodology.

In today physics, mathematical models are playing the decisive role. If you are a theoretical physicist and you have a mathematically consistent model of a given phenomenon for which you predict it could exist, nobody will ask you about the epistemological stability of your model. On the contrary, an experimentalist will try to prove your model with some indirect experiment.

Let us go back to the end of the 18th century and imagine that we are actively participating in the ether research. We cannot see the ether directly with our senses. We can only see the light which we suppose is the wave of the ether. Our idea is that ether is not stationary, ether is moving with the objects. We can imagine the planet Earth moving through the ether in the way that ether which is around the Earth is moving and rotating with the Earth. Ether we cannot strictly divide from the physical object, they are intrinsically bonded and should be examined together. Such imagination is rescuing the ether hypothesis without the introduction of the length contraction. Photon is the wave of ether and behaves according to the Doppler effect. Every inertial system is moving in the ether and light which is the wave of ether has the same velocity in every inertial system. When the distance between the source of the light and the inertial system that is shortening the light will increase frequency, when the distance is increasing, the light frequency will decrease. Ether which is surrounding the Earth is rotating with the Earth. We call this phenomenon ether drift (in today physics it is called "quantum vacuum dragging effect") and it fully explains and mathematically describes Sagnac effect, which STR cannot explain (Fiscaletti and Sorli, 2016).

In physics, it happens that wrong imaginations are summed up. The idea of stationary ether has caused that ether was thrown out of the physics. Having no more the medium of light, Einstein has created the idea that a photon can move in an empty space deprived of physical properties. This idea is the cause of the crisis in today's physics, in which Standard model tries to describe physical reality only with the fields and particles which exist in an empty space. We have, in today physics, three main fields (electromagnetic quantum vacuum of quantum electrodynamics QED, Higgs field and gravity field) and we are not able to recombine these fields in a unique model.

Advanced Relativity model is staying with the ether under the new name "dynamic quantum vacuum", where photon is the wave of quantum vacuum. Inspired by the work of Max Planck, the ether is given a physical property of Planck energy density, which has minimal variations according to the mass of a given physical object: diminished energy density of ether (quantum vacuum) corresponds to the mass of a given physical object:

$$E = mc^2 = (\rho_{PE} - \rho_{qvE}) \cdot V \qquad (2),$$

where ρ_{PE} is Planck energy density, ρ_{qvE} is energy density of quantum vacuum in the centre of a given particle (or massive body), m is mass of the particle (or massive body), V is volume of the particle (or massive body) (Sorli et al.,2017). Advanced Relativity model works perfectly without Higgs field and without gravitational field. Mass and gravity both originate in the variable energy density of quantum vacuum.

Bijective epistemology requires that the phenomenon which is examined needs to be observed with the senses of the observer or detected with an instrument. Human perception and instrumental perception are the obligatory elements for the beginning of the research on a given subject. When the research subject is predicted theoretically, that is, based on an existent model, it needs to be confirmed with a direct human observation or direct instrumental measurement. A good example of this is Dmitri Mendeleev's research, who published the first periodic table of the chemical elements in 1869, based on properties that appeared with some regularity, as he laid out the elements from the lightest to the heaviest (Kaji, 2002). When Mendeleev proposed his periodic table, he noted gaps in the table and predicted that as-then-unknown elements existed with properties appropriate to fill those gaps. Unknown elements have been later discovered by different researchers.

The discoveries of Higgs field and gravitational waves are epistemologically weak, because they have not been measured directly. Today, they are recognized as big achievements of physics. If this will still be the case in 2117, it is doubtful. There is no doubt that the periodic system of elements will be still valid in 2117, because it is based on direct measurements. Indirect measurements applied in the research of Higgs field and gravitational waves should be carefully examined before their full application can be used as the standard in future research. The question seems philosophical, but it is not. It touched the core of physics and deserves attention of both theoretical physicists and experimentalists.

Bijective epistemology is fulfilling Einstein's vision of "completeness" of a theory": "If, without in any way disturbing a system, we can predict with certainty (i.e., with probability equal to unity) the value of a physical quantity, then there exists an element of physical reality corresponding to this physical quantity." And for a theory to be complete, "every element of the physical reality must have a counterpart in the physical theory" (Bernstein 1999). In Advanced Relativity, every element in the model corresponds to exactly one element in the physical reality.

Einstein used to say: "Imagination is more important than knowledge. For knowledge is limited, whereas imagination embraces the entire world, stimulating progress, giving birth to evolution." We must add here that scientific imagination, in order to lead us to coherent models, needs to be based on human perception and experimental data. If not, imagination can be developed in a mathematical model (with no real correspondence with the physical world), for which we then tirelessly search until we confirm it with an indirect measurement. We are proposing a new research methodology in physics, which is fully allowing creative imagination and is based on human perception and experimental data.

Bijective research methodology is excluding the possibility of an error in the process of scientific research in physics. This methodology is giving more credibility to the creative imagination based on perception, rather than to pure mathematic speculation, which is often disconnected from the physical world. Higgs mechanism, for example, is based on pure mathematics speculation and is as such epistemologically unstable.

Advanced Relativity is based on bijective research methodology. Advanced Relativity has kept the ether as the physical basis of the universal space. Ether is not in the space; ether is the "stuff" out of which space is made. Ether is dynamic in the sense that ether is moving with the objects and is rotating with them. We call ether with the new name "quantum vacuum". In the end of the 18th century, light was understood as the wave of the ether. We cannot see and detect ether directly; we can see and detect light as its vibration. Advanced Relativity has adopted this view because Einstein's idea that photon could move in an empty space deprived of physical properties is more philosophical than scientific: light needs a medium and this medium must also be an element of the scientific model of the physical world. Between ether (quantum vacuum) of physical reality and the ether (quantum vacuum in the model of Advanced Relativity) there is a bijective function:

$$f : X \rightarrow Y \quad (1),$$

where X represents ether in physical reality and Y represents ether in the model of Advanced Relativity.

Advanced Relativity is built on bijective methodology and has following advances:

1. In the universal space, there is always NOW. Linear time belongs to the mind. With clocks, we measure the duration of material changes, i.e. motions in space.

2. No signal can move in time, every signal can move in space only and time is the duration of its motion. CMBR radiation is the radiation of quantum vacuum, where there is always NOW. Big Band Theory has come to the end.

3. In the space velocity of light, it has minimal changes and depends on the energy density of quantum vacuum. "Gravitational time dilation" actually means that light minimally diminishes the speed, because in stronger gravity, energy density of quantum vacuum

is a bit lower, which changes its permittivity and permeability, and so velocity of light. We know from classical physics that density of the medium defines velocity of the signal. The denser the medium, the faster the motion of the signal. This is also valid in Advanced Relativity.

4. Mass and gravity both have origin in variable energy density of quantum vacuum.

5. Dark energy is the energy of quantum vacuum.

6. In Advanced Relativity, there is no "inner observer", "outer observer", "coordinate time", "proper time". Velocity of clocks in all inertial systems is valid for all observers and does not depend on them.

7. Relative velocity of clocks in inertial systems depends only on the variable energy density of quantum vacuum and is valid for all observers. GPS proves that clocks on the satellites and on the Earth surface run with the same velocity for all observers. It if would

not have been be so, GPS could not work *(Fiscaletti and Sorli, 2016).*

8. *Curvature of space in GRT is the mathematical description of variable energy density of quantum vacuum. Space is not "curved" in a physical sense (Fiscaletti and Sorli, 2014). NASA confirms universal space is "flat"; it has Euclidean form (NASA, 2013).*

Epistemology crisis of today physics has roots in the conviction that the development of new mathematical models will develop physics. This is true only partially because mathematical models of today physics are often the result of pure speculation, which is not grounded in human perception and experimental data. To overcome the crisis, this article has proposed a new methodology, which allows only imagination based on human perception and experimental data.

References:
Bernstein, H.J. Foundations of Physics (1999) 29: 521.
https://doi.org/10.1023/A:1018856024112

Fiscaletti D. and Sorli A., SPACE-TIME CURVATURE OF GENERAL RELATIVITY AND ENERGY DENSITY OF THERE DIMENSIONAL QUANTUM VACUUM, Annales UMCS Sectio AAA: Physica, VOL.LXV. (2014).

Fiscaletti D and Sorli A., Bijective epistemology and space-time. Foundations of Science; 20(4): 387-398. (2015)

Fiscaletti D. and Sorli A., Dynamic Quantum Vacuum and Relativity. Annales UMCS Sectio AAA: Physica, VOL. LXXI. (2016).

FitzGerald, George Francis (1889), "The Ether and the Earth's Atmosphere", Science, 13 (328): 390.

Kaji, Masanori (2002). "D. I. Mendeleev's concept of chemical elements and The Principles of Chemistry". Bulletin for the History of Chemistry. **27** (1): 4–16.

NASA, http://map.gsfc.nasa.gov/universe/uni_shape.html (2013).

Sorli A., Kaufman S., Dobnikar U., Fiscaletti D., Advanced Relativity for the Renaissance of Cosmology and Evolution of Life, NeuroQuantology (in press).

9.   What is the origin of true human intelligence?

True intelligence is the capacity to have an overview over our thinking. Thinking without awareness leads most of the time to disaster. When we are aware of our thinking it means that we are in contact with consciousness. When the human mind is thinking on its own it creates chaos. How's that? The fact is that in the universe, all elements are interconnected through the consciousness. When the human mind is not integrated with consciousness it functions on its own as if the rest of the universe does not exist. These egoic actions bring suffering to the person who is acting and to the others involved. When our minds are connected with consciousness, they function in harmony with other people, nature and the entire universe.

The whole process of education worldwide is to build the individuality of the person. Interconnectedness is not the subject of education. The result is a new liberalism life-philosophy, which creates a global economic and cultural crisis. The only solution is that we integrate meditation in schools worldwide. True intelligence, which supports life and builds harmony, has its origin in consciousness.

"Artificial intelligence" is the wrong term. Computers are and will remain in the domain of "logic" because they are limited with the 3D physical world. We can only talk about "artificial logic". Intelligence belongs to living beings and has origin in the cognitive activity of higher dimensional Hilbert spaces. TRUTH is the real source of intelligence.

10. Conscious communication for Intercultural Citizenship

Physics of TRUTH is advocating "The Intercultural Citizenship Education" initiative of Anna Lindh Foundation. People from different cultures have different education and so different mindsets through which they perceive and experience life. In order to build an intercultural society, we need to develop clear communication which is free of prejudices and stereotypes of different cultures. It is not enough that we learn about different cultures in order to see clearly a given situation and to have clear communication we have to step out of our own cultural background.

Intercultural society needs a new toll which will make possible that we all take a step away and see our lives from a given distance which makes us free, not only from our culture, but also from all other cultures' prejudices and stereotypes. When we see our own culture and all other cultures from a given distance more space is created and this space gives us the possibility to develop deep and clear communication.

Physics has a tool which can be implemented in the development of an intercultural society; this tool is what we call in physics "the observer". The observer has the inherent ability to observe outer physical phenomena and inner psychological phenomena. The development of the observer has not been implemented yet in schools worldwide. The human hidden capacity to become fully aware of the way its mind works has not been discovered in western science yet. Recent research of our research group has shown that the observer and its capacity to become aware the way a scientific mind works is of immense importance when building a new scientific paradigm, free of some of the prejudices of science in the 20th century. The observer can also be fully engaged in the development of an intercultural society; it gives each individual the possibility of becoming aware of his own prejudices, the prejudices of other cultures and so to develop clear and deep communication with people from another cultures.

Our research confirms that the observer is the function of the same entity which is consciousness itself. In each human being, regardless of their religion, nationality, race and geographic origin, act with the same consciousness as the observer. On this scientific discovery we can build a powerful tool with clear and deep communication which will reach into the core of a human being where we are all one consciousness. The discovery of consciousness in each individual worldwide is the corner-stone of intercultural society. Conscious communication is the most useful tool for intercultural dialogue and the growth of an intercultural society.

Witnessing Method (WM) to discover the TRUTH

The Witnessing Method is an extremely efficient method to get to know the TRUTH. It is much more efficient than any kind of meditation. First you observe (witness) the matter around you. Second you witness your mind's activity. Third you witness the space in which the mind exists. The TRUTH is beyond the space.

11. Open letter to the scientific community

Science without consciousness does not serve the real needs of today humanity. Science with consciousness means the activation of the observer so that he is able to see how a scientific mind works. This is the core research of "cognitive science". The conscious observer has their roots in consciousness itself. Consciousness is subjective and cannot be searched as an object. Consciousness is beyond matter, mind and the space in which matter and mind exist. Consciousness functions as the pure witnessing of which physical and mental objects exist.

In today's science the "conscious observer" has not yet awakened. The result is that today's science is lost in its own reductionist paradigm. It does not produce any good results for human beings. The purpose of today's science is to perpetuate research to get a monthly salary with no benefit for humanity. This is true for human as well as for natural sciences.

We have on this planet many peace institutes. I have not found a single article which gives us a clear model of how to develop peace on this globe. What these "peace scientists" search for the good of humanity and waste public money, is a good question for sociologists who have not understood yet that human society cannot be approached in a proper way without being placed in the universe. We humans live in the universe. Sociology has not achieved that yet.

We have physicists which are looking inside the nucleus of the atom and they are not aware who is doing that. They believe, the Standard model is an adequate picture of physical reality and to be even more convinced they hand out a Nobel prize given for the discovery of the Higgs mechanism which is pure scientific fantasy. No Higgs field exists in the physical universe. It is all "made up" in order to continue "research" with more powerful cyclotron's which will give us new "particles". There is only one "particle" in the universe which is a proton and has inertial mass which is the necessary physical property for a given physical object to be called "particle". Quarks and leptons are different vortexes of a quantum vacuum. A gluon and photon are also vortexes of the quantum vacuum. The Z boson and W boson are fluxes of the quantum vacuum released by the weak interaction. They have their existence in a physical universe on their own. They are not manmade. Meanwhile the Higgs boson is a manmade flux of quantum vacuum released by the relativistic collisions of protons. It has no existence in the universe on its own. Saying that the Z boson, W boson and Higgs boson are "particles" is wrong, because they do not have inertial mass and they are unstable.

Today science is in a deep crisis which we can only overcome by awakening the conscious observer who is fully aware how the scientific mind works. The scientific mind without consciousness is totally lost in its own paradigm and he is fooling himself that he will be able to "adjust" the model in the future and everything will be fixed. We have now in physics three fields: the electromagnetic quantum vacuum of QED, gravity field and Higgs field. Nobody is able to put them together, simply because the gravity field and Higgs field are pure "scientific illusions" with no physical existence. Physicists are comforting themselves that once in the future these three fields will be reconnected in one model. This will never happen; it is another illusion invented for keeping them calm and that "all is right". It is not right, physics loses the ground under their feet and without the introduction of bijective epistemology there will be no progress.

Dear scientist, start actively witnessing your scientific mind "making" its science and in a few weeks you will discover you are completely trapped in your own mind. The only step out is to deepen your ability to become a conscious observer. Only the conscious observer has the power to step out of the old scientific paradigm and build a new one which will serve human beings.

You are the observer, the witness of space in which mental and physical phenomena appear. You are beyond matter, space, mind, and time. Forget about the contents in space, be aware of space itself and you will enter the TRUTH.

Your perception is the shortcut into TRUTH. Look at your hand. Close your eyes and observe the mental image of your hand. Observe outer space in which is your physical hand; observe the inner space in which is the imagine of your hand. It is the same space. You are a witness to this space in which the whole universe exists.

Practise this simple exercise every day and it will change your life completely. You do not need to believe in God, you do not need to be an atheist, just be aware who is witnessing the space and you will take the plunge into the TRUTH. This is the scientific path into TRUTH which is not built on any belief or theory. It is based exclusively on your perception, which is the fundamental source of your knowledge and wisdom.

Yours sincerely, Amrit Srecko Sorli

Reader notes:

Made in the USA
Columbia, SC
19 October 2017